ENVIRONMENTAL PROTECTION ACT 1990

SECTION 34

WASTE MANAGEMENT

THE DUTY OF CARE

A CODE OF PRACTICE

March 1996

London: HMSO

© Crown copyright 1996

Applications for reproduction should be made to HMSO's Copyright Unit, St Crispins, Duke Street, Norwich, NR3 1PD

ISBN 0 11 753210 X

CONTENTS

                                                            Page

Introduction	3
THE CODE OF PRACTICE	
STEP BY STEP GUIDANCE	
Identify and describe the waste	6
Keep the waste safely	11
Transfer to the right person	13
Receiving waste	20
Checking up	23
Expert help and advice	27
The duty of care and scrap metal	30
SUMMARY CHECKLIST	37
ANNEXES	
A. The law on the duty of care	39
B. Responsibilities under the duty of care	47
C. Regulations on keeping records	51
D. Other legal controls	54
i. Waste management licensing	54
ii. The registration of waste carriers	55
iii. The registration of brokers	56
iv. Special waste	57
v. Road transport of dangerous substances	57
vi. International waste transfers	58
vii. Health and safety	58
E. Glossary of terms used in this code of practice	60
Appendix: The definition of waste	64

INTRODUCTION

i. This code of practice consists of the guidance in Sections 1-7 together with their related Annexes. It is issued by the Secretary of State for the Environment, the Secretary of State for Scotland and the Secretary of State for Wales in accordance with section 34(7) and (8) of the Environmental Protection Act 1990 ("the 1990 Act"). This code supersedes that issued in December 1991 which is hereby revoked. This introduction is not part of the code of practice.

ii. Section 34 of the 1990 Act imposes a duty of care on persons concerned with **controlled waste**. The duty applies to any person who produces, imports, carries, keeps, treats or disposes of controlled waste, or as a broker has control of such waste. Breach of the duty of care is an offence, with a penalty of an unlimited fine if convicted on indictment.

iii. Waste poses a threat to the environment and to human health if it is not managed properly and recovered or disposed of safely. The duty of care is designed to be an essentially self regulating system which is based on good business practice. It places a duty on anyone who in any way has a responsibility for controlled waste to ensure that it is managed properly and recovered or disposed of safely. The purpose of this code is to provide practical guidance for waste holders and brokers subject to the duty of care. It has been revised to take account of changes in waste management law since the code was first issued in December 1991. These changes include the application of the duty of care to the scrap metal industry from 1 October 1995.

iv. The code recommends a series of steps which should normally be enough to meet the duty. The code cannot cover every contingency. The legal obligation is to comply with the duty of care itself rather than with the code. Annex A gives a detailed explanation of the law.

v. The Code makes reference to the Environment Agency and to the Scottish Environment Protection Agency. In England and Wales, local authorities' waste regulation functions are transferred to the Environment Agency by section 2 of the Environment Act 1995 with effect from 1 April 1996. Until that date, waste regulation is a function of the local authorities designated in section 30(1) of the Environmental Protection Act 1990.

vi. In Scotland, local authorities' waste regulation functions are transferred to the Scottish Environment Protection Agency by section 21 of the Environment Act 1995 with effect from 1 April 1996.

vii   The code is divided into:-

> - Sections 1-7: Step by step guidance on following the duty;
> - A summary check list; and
>
> Annexes   A: the law on the duty of care;
> B: responsibilities under the duty;
> C: Regulations on keeping records;
> D: an outline of some other legal requirements; and
> E: a glossary of terms used in the code.

viii. Guidance on the definition of waste, which does not form part of the code of practice, is contained in the Appendix.

Is it "waste"?

ix. The duty of care applies to waste which is controlled waste (see paragraphs 1.1-1.2 below). The first question, therefore, is whether any particular substance is "waste". Since 1 May 1994 the common European definition of waste in the Framework Directive[1] on waste has been in force. The legal definition of waste is:-

> "any substance or object...which the producer[2] or the person in possession of it discards or intends or is required to discard."

x. Whether or not a substance is waste must be determined on the facts of the case; and interpretation of the law is a matter for the Courts. The Departments[3] have provided guidance on the interpretation of the definition of waste in DOE Circular 11/94[4]. In the Departments' view the purpose of the Framework Directive is to treat as waste, and accordingly to supervise the collection, transport, storage, recovery and disposal of, those substances or objects which fall out of the commercial cycle or out of the chain of utility. The Departments have suggested, therefore,

---

[1] Council Directive 75/442/EEC as amended by Directives 91/156/EEC and 91/692/EEC.

[2] "Producer" means anyone whose activities produce waste or who carries out pre-processing, mixing or other operations resulting in a change in its nature or composition.

[3] The Department of the Environment, the Welsh Office and The Scottish Office Agriculture, Environment and Fisheries Department.

[4] Welsh Office Circular 26/94 and The Scottish Office Environment Department Circular 10/94.

that to determine whether a substance or object has been discarded the following question may be asked:-

> Has the substance or object been discarded so that it is no longer part of the normal commercial cycle or chain of utility?

xi. An answer of "no" to this question should provide a reasonable indication that the substance or object concerned is not waste. As indicated, however, the purpose of the Framework Directive is to supervise the collection, transport, storage, recovery and disposal of waste; and these are activities which themselves may be of a commercial nature. A distinction must be drawn, therefore, between the normal commercial cycle and the commercial cycle which exists for the purpose of collecting, transporting, storing, recovering and disposing of waste. It is also essential to bear in mind that a substance or object does not cease to be waste as soon as it is transferred for collection, transport, storage, recovery or disposal. A substance or object which is waste at the point of its original production should be regarded as waste until it is recovered or disposed of. **The Appendix** provides a summary of the main questions which need to be addressed in reaching a view on whether a particular substance or object is waste; and when it may cease to be waste.

STEP BY STEP GUIDANCE

SECTION 1

IDENTIFY AND DESCRIBE THE CONTROLLED WASTE

1.1 The duty of care applies to anyone who is the holder[5] of controlled waste[6]. The only exception to this is for the occupiers of domestic property for the household waste which comes from their home. Anyone subject to the duty of care who has some "controlled waste" must identify and describe the kind of waste it is.

Is it "controlled waste"?

1.2 "Controlled waste" means waste from households[7], commerce or industry. At present, the main kinds of waste that are not "controlled waste" are waste from agricultural premises, waste from mines and quarries, explosives and most radioactive waste. However, the Departments intend to issue a consultation paper containing proposals for the extension of the definition of controlled waste to include certain categories of agricultural waste and of mines and quarries waste. Unless the context indicates otherwise, subsequent references in the code to "waste" should be read as references to "controlled waste".

What are the problems of the waste?

1.3 Waste cannot be simply divided between the safe and the hazardous. There are safe ways of dealing with any waste. Equally, any waste can be hazardous to human health or the environment if it is wrongly managed. Deciding whether any waste poses a problem requires consideration not only of its composition but of what will happen to it. For most waste it is not necessary to know more than what it is in very general terms. But subsequent holders must be provided with a description of the waste that is full enough to enable them to manage the waste properly. Even everyday items may cause problems in handling or treatment.

---

[5] "Holder" means a person who imports, produces, carries, keeps, treats, or disposes of controlled waste or, as a broker, has control of it.

[6] Only waste which is "Directive waste" is treated as controlled waste - see the glossary of terms at **Annex E**.

[7] Householders are exempt from the duty of care for their own household waste (**Annex A** paragraph A.9).

1.4 In looking for waste problems it may help to ask such questions as:-

   (a) does the waste need a special container to prevent its escape or to protect it from the elements;

   (b) what type of container suits it and what material can the container be made of;

   (c) can it safely be mixed with any other waste or are there wastes with which it should not be mixed;

   (d) can it safely be crushed and transferred from one vehicle to another;

   (e) can it safely be incinerated or are there special requirements for its incineration, such as minimum temperature and combustion time;

   (f) can it be disposed of safely in a landfill site with other waste; and

   (g) is it likely to change its physical state during storage or transport?

1.5 Anything unusual in waste can pose a problem. So can anything which is out of proportion. Ordinary household waste, and waste from shops or offices, often contains small amounts of potentially hazardous substances. This may not matter if they are mixed in a large quantity of other waste. What should be identified as potential problems in a consignment of waste, are significant quantities of an unexpected substance, or unusual amounts of an expected substance.

1.6 Note that certain particularly dangerous or difficult wastes are subject to strict legal controls quite apart from and additional to the requirements of the duty of care. These wastes are known as "special wastes" (see **Annex D** paragraphs D.16-D.18).

What goes in a description?
1.7 A transfer note must be completed, signed and kept by the parties involved if waste is transferred. This is a requirement of the Environmental Protection (Duty of Care) Regulations

1991 ("the 1991 Regulations"). Any breach of the 1991 Regulations is an offence (see **Annex C** on keeping records). Amongst other things, the transfer note must state:-

   (a) the quantity of waste transferred - most waste management companies, in particular landfill operators, receive quantities of waste by weight. Wherever possible, therefore, a transfer note should record the weight of waste transferred;

   (b) how it is packed - whether loose or in a container; and

   (c) if in a container, the kind of container.

1.8 There must also be a description of the waste. This may be provided separately or combined as a single document with the transfer note. At the time of the publication of this code, a national waste classification scheme is the subject of consultation. One of the aims of the scheme is to provide a comprehensive list of waste types which may be generally accepted as the reference point for describing waste. The description of waste should, therefore, wherever possible refer to the appropriate entry in the national waste classification scheme when, subject to consultation, it is introduced. It is good practice to label drums or similar closed containers with a description of the waste. Under the 1991 Regulations the parties must keep the transfer note and the description for two years. The description should always mention any **special problems**, requirements or knowledge. In addition it should include some combination of:-

   (a) the type of premises or business from which the waste comes;

   (b) the name of the substance or substances;

   (c) the process that produced the waste; and

   (d) a chemical and physical analysis.

1.9 The description must provide enough information to enable subsequent holders to avoid mismanaging the waste. When writing a description it is open to the holder to ask the manager who will handle the waste what he needs to know. For most wastes, those that need only a simple description, either 1.8(a) or (b) above will do. However, in some cases it may not be enough simply to describe the waste as "household", "commercial" or "industrial" waste without providing a clearer idea of what the substance is or providing details about the premises from which it originated. Each element of the description is dealt with in paragraphs 1.11-1.17 below.

Special problems

1.10 The description should always contain any information which might affect the handling of the waste. This should include:-

(a) special problems identified under paragraphs 1.3 to 1.6 above;

(b) any information, advice or instructions about the handling, recovery or disposal of the waste given to the holder by the Environment Agency or by the Scottish Environment Protection Agency [8] or the suppliers of material or equipment;

(c) details of problems previously encountered with the waste;

(d) changes to the description since a previous load.

A) **Source of the waste: type of premises or business**

1.11 It may sometimes be enough to describe the source of the waste by referring either to the use of the premises where the waste is produced or to the occupation of the waste producer.

1.12 Such a "source of waste" description is recommended as the commonsense simple description where businesses produce a mixture of wastes none of which has special handling or disposal requirements; or where there are no special handling or disposal requirements which cannot be identified from such a simple description. Such a description must make it clear what type of wastes are produced and the contents of the wastes; and their proportions must be only such as might be expected.

B) **Name of the substance**

1.13 The waste may be described by saying what it is made of. This may be in physical and chemical terms or by the common name of the waste where this is equally helpful. Such a

---

[8] In the case of processes regulated under Integrated Pollution Control in England and Wales, the Environment Agency may give advice to the producer of waste that is relevant to its subsequent management prior to its final disposal or treatment. In Scotland the Integrated Pollution control authority is the Scottish Environment Protection Agency. This advice should also be included in the description. .

description by name is recommended for waste composed of a single simple material or a simple mixture.

C) Process producing the waste

1.14 The waste may be described by saying how it was produced. Such a description would include details of materials used or processed, the equipment used and the treatment and changes that produced the waste. If necessary this would include information obtained from the supplier of the materials and equipment.

1.15 This should form part of the description for most industrial wastes and some commercial wastes.

D) Chemical and physical analysis

1.16 A description based on the process producing the waste (C) will not go far enough where the holder does not know enough about the source of the waste. It will often not be adequate where:-

(a) wastes, especially industrial wastes, from different activities or processes are mixed; or

(b) the activity or process alters the properties or composition of the materials put in.

1.17 For such wastes an analysis will usually be needed. In cases of doubt, the holder may find it helpful to consult the intended waste manager as to whether he needs an analysis to manage the waste properly. Where it is necessary the holder should detail the physical and chemical composition of the waste itself including, where different substances are mixed, their dilutions or proportions. The holder might either provide this information himself or obtain a physical or chemical analysis from a laboratory or from a waste management contractor.

SECTION 2

KEEP THE WASTE SAFELY

The problem

2.1 All waste holders must act to keep waste safe against:-

(a) corrosion or wear of waste containers;

(b) accidental spilling or leaking or inadvertent leaching from waste unprotected from rainfall;

(c) accident or weather breaking contained waste open and allowing it to escape;

(d) waste blowing away or falling while stored or transported;

(e) scavenging of waste by vandals, thieves, children, trespassers or animals.

2.2 Holders should protect waste against these risks while it is in their possession. They should also protect it for its future handling requirements. Waste should reach not only its next holder but a licensed facility or other appropriate destination without escape. Where waste is to be mixed immediately, for example in a transfer station, a civic amenity site or a municipal collection vehicle, it only needs to be packed well enough to reach that immediate destination. Preventing its escape after that stage is up to the next holder. However, there are wastes that may need to reach a disposal or treatment site in their original containers. For example, drummed waste. In such cases, holders will need to know through how many subsequent hands; under what conditions; for how long; and to what ultimate treatment their waste will go in order to satisfy themselves that it is packed securely enough to reach its final destination intact. If an intermediate holder alters waste in any way, by mixing, treating or repacking it, then he will be responsible for observing all this guidance on keeping waste safe.

Storing waste securely

2.3 Security precautions at sites where waste is stored should prevent theft, vandalism or scavenging of waste. Holders should take particular care to secure waste material attractive to scavengers, for example, building and demolition materials and scrap metal. Special care should also be taken to secure waste which has a serious risk attached to it, for example certain types of clinical waste. Waste holders should undertake regular reviews of the waste in their possession to ensure that it has not been disturbed or tampered with.

2.4 Segregation of different categories of waste where they are produced may be necessary to prevent the mixing of incompatible wastes. For example, avoiding reactions in mixtures. Segregation may assist the disposal of waste to specialist outlets. Where segregation is practised on sites, the waste holder should ensure that his employees and anyone else handling waste there are aware of the locations and uses of each segregated waste container.

Containers

2.5 Waste handed over to another person should be in some sort of container, which might include a skip. The only reasonable exception would be loose material loaded into a vehicle and then covered sufficiently to prevent escape before being moved. Waste containers should suit the material put in them.

2.6 It is good practice to label drums or similar closed containers with a note of the contents when stored or handed over. This could be a copy of the waste description. To avoid confusion, old labels should be removed from drums which are reused.

Waste left for collection

2.7 Waste left for collection outside premises should be in containers that are strong and secure enough to resist not only wind and rain but also animal disturbance, especially for food waste. All containers left outside for collection will therefore need to be secured or sealed. For example, drums with lids, bags tied up, skips covered. To minimise the risks, waste should not be left outside for collection longer than is necessary. Waste should only be put out for collection on or near the advertised collection times.

SECTION 3

TRANSFER TO THE RIGHT PERSON

3.1 Waste may be handed on only to authorised persons or to persons for authorised transport purposes. This **Section** of the code advises on who these persons are; and what checks to carry out *before* making an arrangement or contract for transferring waste.

3.2 Section 34(3) of the 1990 Act sets out those who are authorised persons. The list of authorised persons is:-

(a) any authority which is a waste collection authority for the purposes of Part II of the 1990 Act;

(b) any person who is the holder of a waste management licence under section 35 of the 1990 Act;

(c) any person who does not need a waste management licence because the activity he carries out is exempt from licensing by virtue of regulation 16 or 17 of and Schedule 3 to the Waste Management Licensing Regulations 1994[9] ("the 1994 Regulations");

(d) any person who is registered as a carrier of controlled waste under section 2 of the Control of Pollution (Amendment) Act 1989 ("the 1989 Act");

(e) any person exempt from registration as a carrier of controlled waste under regulation 2 of the Controlled Waste (Registration of Carriers and Seizure of Vehicles) Regulations 1991[10]; and

(f) In Scotland, a waste disposal authority acting in accordance with a resolution made under section 54 of the Environmental Protection Act 1990 (see paragraphs 3.20-3.21).

3.3 The list of authorised transport purposes is set out in section 34(4) of the 1990 Act which is reproduced at the end of **Annex A** (see page 44).

---

[9] As amended by the Waste Management Licensing (Amendment etc.) Regulations 1995.

[10] S.I. 1991 No.1624 as amended by the Controlled Waste Regulations 1992 (S.I. 1992 No 588) and the Waste Management Licensing Regulations 1994 (S.I. 1994 No 1056).

3.4 It should be noted that it is an offence under Regulation 20 of the 1994 Regulations (subject to various provisions) for anyone to arrange on behalf of another person for the disposal or recovery of controlled waste if he is not a registered broker. Anyone subject to the duty of care must ensure that insofar as they use a broker when transferring waste, they use a registered broker (that is any person registered as a waste broker in accordance with regulation 20 of the 1994 Regulations) or one exempt from the registration requirements (in accordance with regulation 20 of the 1994 Regulations) (see Annex paragraphs D.11-D.15).

Public waste collection

3.5 Local authorities collect waste from households and from some commercial premises. They do this either with their own labour or using private contractors who will be registered carriers (**Annex D** paragraphs D.7-D.10). Private persons are exempt from the duty in connection with their own household waste produced on their premises (**Annex A** paragraph A.9). If there is any doubt about whether or not a particular waste can go in the normal collection, the producer should ask the local authority (the borough or district council or in Scotland on 1 April 1996, the council, or in Wales on 1 April 1996, the county or county borough council).

Using a waste carrier

3.6 A waste holder or broker may transfer waste to someone who transports it - a waste carrier - who may or may not also be a waste manager (**Annex D** paragraphs D.7-D.10). Subject to certain exemptions, anyone carrying waste in the course of their business, or in any other way for profit, must be registered with the Environment Agency or with the Scottish Environment Protection Agency. The Agencies' register of carriers is open to public inspection. For the purpose of the duty of care, holders may use these registers as a reference list of carriers who are authorised to transport waste. However, inclusion on the Agencies' register is *not* a recommendation or guarantee of a carrier's suitability to accept any particular type of waste. The holder or broker should remain alert to any sign that the waste may not be legally dealt with by a carrier.

3.7 Anyone intending to transfer waste to a carrier will need to check that the carrier is registered or is exempt from registration. A registered carrier's authority for transporting waste is either his certificate of registration or a copy of his certificate of registration *if it was provided by the Agencies (or before 1 April 1996 by the relevant waste regulation authority)*. The certificate or copy certificate will show the date on which the carrier's registration expires. All copy certificates must be numbered and marked to show that they are copies and have been provided by the Agencies (or before 1 April 1996 by the relevant waste regulation authority). Photocopies are not valid and **do not** provide evidence of the carrier's registration.

3.8 In all cases other than those involving repeated transfers of waste, the holder should ask to see, and should check the details of, the carrier's certificate or copy certificate of registration. In addition, before using any carrier for the first time, the holder should check with the Environment Agency or the Scottish Environment Protection Agency that the carrier's registration is still valid, even if his certificate appears to be current. The holder should provide the appropriate Agency with the carrier's name and registration number as shown on the certificate.

3.9 In practice, the only exempt carriers who might take waste from a holder are:-

(a) charities and voluntary organisations;

(b) waste collection authorities (local authorities) collecting any waste themselves (though an authority's contractors are not exempt); waste disposal authorities and the Agencies;

(c) wholly owned subsidiaries of British Rail when carrying waste by rail[11];

(d) ship operators where waste is to be disposed of under licence at sea;

(e) persons who are authorised under the Animal By-Products Order 1992 to hold or deal with animal waste or the holder of a knackers yard licence[12].

3.10 It should be noted that animal waste which is collected and transported in accordance with Schedule 2 to the Animal By-Products Order 1992 is not controlled waste for the purposes of the duty of care (see **Annex A** paragraph A.10).

---

[11] Regulation 23(2) of the 1994 Regulations amends the controlled Waste (Registration of Carriers and Seizure of Vehicles) Regulations 1991 to provide an exemption from carrier registration for any wholly owned subsidiary of the British Railways Board which has applied for registration as a carrier of controlled waste. The exemption applies only where the subsidiary of the British Railways Board is registered under the provisions of paragraph 12 of Schedule 4 to the 1994 Regulations, and for the period whilst its application for registration under the Controlled Waste Regulations 1991 is pending.

[12] That is, a knackers yard licence which in England and Wales has the same meaning as in Section 34 of the Slaughterhouses Act 1974 and in Scotland means a licence under Section 6 of the Slaughter of Animals (Scotland) Act 1980.

3.11 Charities and voluntary organisations, waste collection, disposal authorities and the Agencies and British Rail subsidiaries need to be registered as exempt carriers in the Agencies' register of professional collectors and transporters of waste[13].

3.12 In all cases other than those involving repeated transfers of waste, the holder should ask the carrier to confirm the type of exemption under which he transports waste. In addition, before using for the first time a carrier who claims to be exempt, the holder should ask the carrier to provide evidence that the exemption which he claims is valid.

Sending waste for disposal, treatment or recovery

3.13 A "waste manager" is anyone:-

(a) who stores waste or who processes it in some intermediate way short of final disposal[14];

(b) who carries out a waste recovery operation[15]; or

(c) who carries out a waste disposal operation[16].

3.14 All of these activities are subject to the waste management licensing requirements of Part II of the 1990 Act. Where a licence is necessary it is issued by the Agencies (and before 1 April 1996 by the relevant waste regulation authority). The licence will usually set out the types and quantities of waste that the operator may deal with and the way in which the waste is managed. It may also cover such matters as operating hours and pollution control on site.

3.15 *Before* choosing a waste manager as the next person to take waste, a holder will need to:-

(a) check that the manager has a licence; and

---

[13] See paragraphs 1.70-1.84 of DOE Circular 11/94 (Welsh Office Circular 26/94 and The Scottish Office Environment Department Circular 10/94).

[14] These activities may involve the deposit of waste under section 33(1)(a) of the 1990 Act or they may be a recovery or disposal operation - see footnotes 14 and 15.

[15] Waste recovery operations are listed in Part IV of Schedule 4 to the 1994 Regulations.

[16] Waste disposal operations are listed in Part III of Schedule 4 to the 1994 Regulations.

(b) establish that the licence permits the manager to take the type and quantity of waste involved.

3.16 A waste holder should check this not merely by asking the waste manager but by examining his licence. The holder in turn should show the manager the description of the waste involved. If the holder doubts whether the licence covers his particular waste he can ask the waste manager or, if he is still not satisfied, the Agency which issued the licence.

3.17 Some forms of waste disposal, treatment or recovery do not require a licence because the activity concerned has been exempted from licensing[17]. If a selected waste manager is not licensed because what he is doing is exempt then he should say so and state which type of exemption he comes under. It is not taking enough care for a holder to consign waste to a contractor who states that he is exempt but does not give the grounds. The holder delivering waste to an exempt waste manager should check that the waste is within the scope of the exemption. The exemptions are limited to specific circumstances and types of waste. If in doubt about the exemption of a particular activity the holder may seek advice from the Agencies. (For more information on licensing and exemptions see **Annex D** paragraphs D.2-D.4).

3.18 A holder of waste should make these same checks on licences and exemptions wherever he delivers waste even if he is not the producer. A carrier should always check that the next holder he delivers to is an authorised person and that the description of the waste he carries is within the licence or exemption of any waste manager to whom he delivers, unless he is only providing the transport to a contract directly between the producer and the waste manager. In that case the producer should make all the checks on the waste manager.

Responsibility of brokers

3.19 A broker, by arranging for the transfer of waste, shares responsibility for its proper transfer with the two holders directly involved. He should follow the relevant part of the guidance in paragraphs 3.6-3.18 above when undertaking checks in connection with arrangements he intends to make, or has made, for waste.

---

[17] By virtue of regulations 16 and 17 of and Schedule 3 to the 1994 Regulations as amended by the Waste Management Licensing (Amendment Etc.) Regulations 1995.

Waste disposal authorities in Scotland, England and Wales

3.20 **In Scotland**, local authorities may operate their own disposal sites for publicly collected waste. In some cases, such sites may also take waste directly from commerce or industry. Sites of this kind are not licensed but are operated subject to the terms and conditions of resolutions passed by the authorities which are broadly similar to those in waste management licences. In relation to these sites, it is necessary to:-

(a) check that they are operated by the local waste disposal authority; and

(b) establish that the resolution permits the authority to take the type and quantity of waste involved.

3.21 The Environment Act 1995 transfers waste regulation functions to the Scottish Environment Protection Agency on 1 April 1996. A period of six months will be allowed from this date for an authority operating a site under a resolution to apply for a waste management licence under Part II of the 1990 Act. Thereafter, the resolution will continue to operate only for so long as it takes to deal with the licence application including the time taken to settle any appeal.

3.22 **In England and Wales**, waste disposal authorities will not normally be responsible for the operation of waste facilities. The Environment Act 1995 nevertheless provides that, from the transfer on 1 April 1996 of waste regulation functions to the Environment Agency, any waste facilities still operated by waste disposal authorities are to be subject to the waste management licensing system under Part II of the 1990 Act.

Checks on repetitive transfers

3.23 Full checks on carriers and waste managers do not need to be repeated if transfers of waste are repetitive - the same type of waste from the same origin to the same destination. The obvious example is waste collection from commercial premises. However, it would be advisable to make occasional checks to ensure that the contents or composition of the waste which is being transferred under cover of a season ticket remains consistent with the waste description.

3.24 For a series of identical loads making up one transaction there is no need to see every time a load of waste is transferred the licence of a waste manager or, in Scotland, the local authority's resolution (see paragraphs 3.20-3.21 above); or the registration certificate of a waste carrier.

However, licences, resolutions, registration or evidence of exemption *should* be examined afresh in the following cases:-

(a) whenever a new transaction is involved, that is if the description or destination of the waste has changed;

(b) where several different carriers or disposers are collecting waste at one place and there might be a danger of an unauthorised carrier collecting a load. For example, a construction or demolition site from which several hauliers are taking waste away;

(c) as a minimum precaution, the licence, resolution, registration or evidence of exemption should be seen and checked at least once a year even if nothing has changed in a series of transfers;

(d) where there has been a change in the licence or resolution conditions of the destination; or

(e) where there has been a change in the waste carrier transporting the waste.

SECTION 4
RECEIVING WASTE

4.1 The previous three **Sections** of this code look at transfers from the point of view of the person transferring the waste to someone else. This **Section** offers guidance to persons receiving waste, whether at its ultimate destination or as an intermediate holder.

Checking on the source of waste

4.2 Checking is not only in one direction. No-one should accept waste from a source that seems to be in breach of the duty of care. Waste may only come *either* from the person who first produces or imports it *or* from someone who has received it. The person who receives it must be one of the persons entitled to receive waste - that is an authorised person or a person for authorised transport purposes (**Section 3**). On the handover of waste, the previous holder and the recipient will complete a transfer note in which the previous holder will declare which category of person entitled to hold waste he is. The recipient should ensure that this is properly completed before accepting waste. Checking back in this way need not be as thorough as checking forward.

4.3 Recipients should not normally need to see any waste management licence of the previous holder; and there is no explicit requirement for a person receiving waste from a waste carrier to check on whether or not the carrier is registered. In the Departments' view, it is not an offence under section 34(1) of the 1990 Act for a recipient to accept waste from an unregistered carrier. However, an offence might have been committed earlier in the chain. For example, a waste producer who transfers waste to someone who is not authorised to accept it. This may be a carrier who was required to be registered but was not. The carrier himself would also be in breach of the carrier registration legislation if he was required to be registered but was not. Paragraphs 5.10-5.12 below provide guidance on reporting to the Agencies.

4.4 However, the first time a carrier delivers waste, the recipient should satisfy himself that he is dealing with someone who is properly registered to transport the waste. In this case, it would be reasonable to ask to see either the carrier's registration certificate or official copy certificate, or to request confirmation of the type of exemption under which the carrier is transporting the waste.

4.5 Before receiving any waste, a holder should establish that it is contained in a manner suitable for its subsequent handling and final disposal or recovery. Where the recipient provides containers he should advise what waste may be placed in them.

4.6 The recipient should also look at the description and seek more information from the previous holder if this is necessary to manage the waste.

Co-operation with the previous holder

4.7 Anyone receiving waste should co-operate with the previous holder in any steps they are taking to comply with the duty. That means in particular supplying correct and adequate information that the previous holder may *need*.

4.8 The previous holder needs to know enough about the later handling of his waste - how it is likely to be carried, stored and treated - to pack and describe it properly. The recipient should give such information.

4.9 Under the 1991 Regulations (**Annex C**) anyone receiving waste must receive a description and complete a transfer note. The recipient must declare on the transfer note which category of authorised person he is, with details. Before making any arrangement to receive waste, a waste manager should show to the previous holder his waste management licence or a statement of the type of exemption from licensing under which he is operating; and a carrier should show his certificate of registration, an official copy of his certificate of registration or evidence of the type of exemption from registration under which he transports waste. It would be sensible for every waste management site office to hold a copy of the waste management licence or a statement of the type of exemption; and for every vehicle used by a carrier to carry an official copy of the carrier's certificate of registration or evidence of the type of exemption under which he transports waste. Where an establishment or undertaking is carrying out an exempt activity, the statement of the type of exemption may take the form of a copy of its entry in the Agencies' register of exempt activities[18].

4.10 A broker, by arranging for the transfer of waste, shares responsibility for its proper transfer with the two holders directly involved. To enable the broker and the other parties to the transfer

---

[18] Regulation 18(8) of the 1994 Regulations requires the Agencies to provide reasonable facilities for obtaining, on payment of reasonable charges, copies of entries in their register of exempt activities.

to discharge their duty of care, information may need to be exchanged between the broker and the previous holder or recipient of the waste, in line with the guidance in this **Section**.

SECTION 5
CHECKING UP

5.1 The previous **Sections** describe normal procedures for transferring waste from the producer to its final destination. Most of the checking that is reasonable is already built into these procedures and the transfer note system (**Annex C**). This **Section** gives guidance on what further checks are advisable and the action to take when checks show that something is wrong.

Checks after transfer

5.2 Most waste transfers require no further action from the person transferring waste after the waste has been transferred. A producer is under no specific duty to audit his waste's final destination. However, undertaking such an audit and subsequent periodic site visits would be a prudent means of protecting his position by being able to demonstrate the steps he had taken to prevent illegal treatment of his waste.

5.3 One exception is where a holder makes arrangements with more than one party. For example, a producer arranges two contracts, one for disposal and another for transport to the disposal site. In that case the producer should establish that he not only handed the waste to the carrier but that it reached the disposer. Similar considerations apply to a broker who may make arrangements on behalf of a waste producer with several parties (carrier, disposer etc).

Checks after receipt

5.4 Any waste holder, but especially a waste manager receiving waste, has a strong practical interest in the description being correct and containing adequate information. A waste management licence will control the quantities and types of waste that may be accepted and how they may be managed. It is within his capacity as a waste manager to ensure that descriptions of waste received are indeed correct. Anyone receiving waste should make at least a quick visual check that it appears to match the description. For a waste manager it would be good practice to go beyond this by fully checking the composition of samples of waste received.

Causes for concern

5.5 Every waste holder should be alert for any evidence that suggests that the duty of care is not being observed or that illegal waste handling is taking place. Obvious causes for concern that any holder should notice when he accepts or transfers waste include:-

(a) waste that is wrongly or inadequately described being delivered to a waste management site;

(b) waste being delivered or taken away without proper packing so that it is likely to escape;

(c) failure of the person delivering or taking waste to complete a transfer note properly, or an apparent falsehood on the transfer note;

(d) an unsupported claim of exemption from licensing or registration as a carrier; and

(e) failure of waste consigned via a carrier to arrive at a destination with whom the transferring holder has an arrangement; or

(f) damage to, or interference with containers.

5.6 Other causes of concern may come to light. Waste holders do not *have* to check up after waste is handed on, but they may become aware of where it is going, and should act on such information if it suggests illegal or careless waste management. Similarly, a broker who suspects mismanagement of waste for which he has made arrangements should take appropriate action. For instance, he may need to make further checks to establish the facts, or use different waste carriers and managers unless and until problems are remedied. Paragraphs 5.10-5.12 give guidance on informing the Environment Agency or the Scottish Environment Protection Agency.

Action to take with other holders

5.7 A holder who only suspects that his waste is not being dealt with properly should first of all check his facts, in the first place with the next (or sometimes the previous) holder. This may involve asking that further details should be added to a waste description, for more information about the exact status of the holder for the purposes of waste management licensing or carrier registration, why they may be entitled to an exemption, or simply where waste went to or came from.

5.8 If a holder is not satisfied with the information or is certain that waste he handles is being wrongly managed by another person then his first action should normally be to refuse to transfer or accept further consignments of the waste in question to or from that person, unless and until the problem is remedied. Such a step may not be practicable in all cases. For example, to avoid breach of a contract to deliver or accept waste or because there is no other outlet immediately available for the waste. Steps should be taken to minimise such inflexibility. One possible measure would be for new waste contracts to provide for termination if a breach of the duty occurs and a notice to rectify is not complied with.

5.9 If an arrangement that has proved wrong in some respect has to continue temporarily then the holder should take stringent precautions. For example, where waste has been misdescribed he should analyse further consignments; where waste has been collected or delivered without being properly packed he should inspect each further load; and where it has not reached its legitimate destination he should check that each subsequent load arrives at the appointed place. He should also, if appropriate, bring the situation to the attention of the Agencies, which may be able to offer guidance.

Reporting to the Environment Agency or to the Scottish Environment Protection Agency
5.10 The Environment Agency and the Scottish Environment Protection Agency are responsible for the licensing of waste management facilities; the registration of establishments or undertakings carrying out activities exempt from licensing; the registration of waste carriers; and the registration of waste brokers.

5.11 The Agencies do not have a specific duty to enforce the duty of care. However, they have a major interest in breaches of the duty which might contribute to illegal waste management. They are also equipped with the powers and expertise to prevent or pursue offences and to advise on the legal and environmentally sound management of waste. Fly tipping (ie disposing of waste without a waste management licence or an exemption from licensing) is one such offence. Fly tipping is a serious offence, and the Courts may impose fines and/or imprison those responsible. The Government is keen to ensure that fly-tipping does not increase following the introduction of the landfill tax in October 1996. The Departments therefore expect the Agencies to give a high priority to the prevention of fly-tipping and to draw the financial gains made by perpetrators to the attention of the Courts.

5.12 Any person who imports, produces, carries, keeps, treats or disposes of waste or who has control of it as a broker is required to take all reasonable measures to prevent the unlawful deposit, recovery or disposal of the waste by themselves or another person. Holders should tell the Agencies where they know or suspect that:-

(a) there is a breach of the duty of care; or

(b) waste is carried by an unregistered carrier not entitled to exemption; or

(c) waste is stored, disposed of, treated or recovered:-

(i) without a licence or in a way not permitted by the licence;

(ii) contrary to the terms or conditions of a licence exemption; or

(iii) in Scotland, contrary to the terms and conditions of an authority's resolution.

SECTION 6
EXPERT HELP AND GUIDANCE

6.1 A waste holder may not always have the knowledge or expertise to discharge his duty of care. He should then seek expert help and guidance. The holder is still the person responsible for discharging his own duty of care, a responsibility that cannot be transferred to an expert adviser.

Help with analysing, describing and handling waste

6.2 Waste consultants and waste managers offer services to examine waste problems, identify wastes and recommend storage and handling methods. A waste holder may also use an analytical laboratory to establish the nature of unknown waste that he needs to describe. A consultant or laboratory acting in this way can only advise, and cannot take over the duty of care from the holder. However reliable the expert, the holder himself needs to ask the right questions. Where he is faced with an unknown substance or a known waste but no disposal outlet, he needs to establish not only what it is but what special needs it has for storage, transport or treatment, and what the possible outlets for it may be, whether disposal or reclamation.

Help with choosing a destination

6.3 A consultant may put a waste holder in touch with a legitimate outlet for his waste. Alternatively, waste managers or carriers themselves may offer expert advice on a destination for a waste. A large waste management organisation with a variety of disposal or recovery methods under its control may be able to provide all the waste management services a holder needs. Otherwise it may be necessary to ask several firms for advice on whether or not they would accept a waste. If a consultant arranges a waste transfer to such an extent that he controls what happens to the waste, he is a broker, and shares responsibility with the two holders directly involved for the proper transfer of the waste (**Section 3** paragraph 3.19 and **Annex B** paragraph B.12).

6.4 If the next holder is himself a source of advice he may sometimes undertake the analysis and write the waste description on behalf of the previous holder. What he may not do is to undertake the checks or complete a transfer note for which the previous holder is responsible.

Help from relevant authorities

6.5 A *waste collection authority* will be able to tell a waste holder whether a particular waste may or will be collected by the authority as part of its normal public waste collection.

6.6 *The Environment Agency and the Scottish Environment Protection Agency* have expertise in the management of all types of controlled waste. Although the Agencies will help as far as possible, they are not in a position to offer advice to every waste holder on how he should deal with his waste. Information that *is* held by the Agencies and is available for any enquirer to examine on request is:-

(a) the register which the Agencies[19] are required to maintain under section 64 of the 1990 Act as amended by the 1995 Act. The information on the register includes details of current or recently current waste management licences;[20]

(b) the register which the Agencies are required to maintain under regulation 18 of the 1994 Regulations of establishments or undertakings carrying on exempt activities involving the recovery or disposal of waste;

(c) the register of waste carriers which the Agencies are required to maintain under regulation 3 of the Controlled Waste (Registration of Carriers and Seizure of Vehicles) Regulations 1991;

(d) the register of waste brokers which the Agencies are required to maintain under regulation 20 of and Schedule 5 to the 1994 Regulations; and

(e) the register of professional collectors, transporters, dealers and brokers of waste which the Agencies are required to maintain under paragraph 12 of Part I of Schedule 4 to the 1994 Regulations.

---

[19] And waste collection authorities in England and, from 1 April 1996, waste collection authorities in Wales

[20] Regulation 10 of the 1994 Regulations sets out the information which the register must contain.

Waste Management Papers

6.7 Detailed advice on good practice is given in the series of Waste Management Papers issued by the Departments and published by HMSO. These cover subjects including the handling, disposal, treatment and recovery of particular wastes and waste management licensing.

SECTION 7

THE DUTY OF CARE AND SCRAP METAL

7.1 Scrap metal became a controlled waste for the purposes of the duty of care from 1 October 1995[21]. The duty of care and the guidance in the previous **Sections** of this code therefore apply to holders of scrap metal as they do to holders of other types of controlled waste. The purpose of this **Section** is to supplement that guidance and, in so doing, to take account of the distinctive features of scrap metal and the circumstances in which it is recovered.

7.2 Metal recycling sites (MRSs) are classified as recovery operations for the purposes of waste management licensing. In this context, the Secretary of State has advised in DOE Circular 6/95[22] that waste regulation authorities (which from 1 April 1996 should be taken to mean the Environment Agency or the Scottish Environment Protection Agency):-

> "(a) should have regard to the fact that scrap metal recovery and waste motor vehicle dismantling are a source of benefit to the environment and sustainable development;
>
> (b) should strike an appropriate balance between advice and encouragement and regulation and legal enforcement; and
>
> (c) should distinguish between and act proportionately in relation to 'technical' breaches of the Regulations[23], the 1994 Regulations[24] or Part II of the 1990 Act where there is no threat of pollution to the environment or harm to human health; and breaches which do give rise to such a threat. In the former case waste regulation authorities' main aim should be to ensure that the person responsible is made aware of his legal responsibilities and that steps are taken by the authority or the person concerned to prevent the commission of any further 'technical' offences."

7.3 The Secretary of State has also recognised the distinctive nature of MRSs in the guidance on the licensing of these facilities which he has issued in Waste Management Paper No.4A (see

---

[21] Regulations 3(2) and 7(2) of the Controlled Waste Regulation 1992 (S.I. 1992 No.588) as amended by regulation 2(1) of the Waste Management Licensing (Amendment etc.) Regulations 1995 (S.I. 1995 No.288).

[22] Welsh Office Circular 25/95 and The Scottish Office Environment Department Circular 8/95.

[23] The Waste Management Licensing (Amendment etc.) Regulations 1995.

[24] The Waste Management Licensing Regulations 1994.

paragraph 6.7 above). For example, the operator of a MRS is likely to have a sound knowledge of the materials he receives; and unlike many other types of controlled waste, a significant proportion of the scrap metal which he receives will have a positive value both to the person transferring it and the person to whom it is transferred. A consequence of this is that the operator of a MRS is likely to have a direct interest in taking measures "to prevent the escape of waste from his control"[25]. Many MRSs operate under the terms of exemptions from waste management licensing. The terms of these exemptions require that the activity concerned is carried on at a "**secure place** designed or adapted for the recovery of scrap metal or the dismantling of waste motor vehicles...."[26]

The transfer note system

7.4 As explained in paragraphs 1.7-1.9 above, if waste is transferred from one person to another a transfer note must be completed, signed and kept by the parties to the transfer. The format of the transfer note is not prescriptive. However, the transfer note must include the information shown on the suggested standard form for voluntary use at **Annex C**.

7.5 Operators of MRSs are well placed to meet the requirements of the 1991 Regulations because of the records which they are required to keep under the Scrap Metal Dealers Act 1964 ("the 1964 Act").[27] In this respect, section 2 of the 1964 Act requires every scrap metal dealer to keep, at each place occupied by him as a scrap metal store, a book[28] and to enter in the book specified particulars with respect to:-

(a) all scrap metal received at that place, and

(b) all scrap metal either processed at, or despatched from, that place.

---

[25] Section 34(1)(b) of the 1990 Act.

[26] Paragraph 45(1) and (2) of Schedule 3 to the Waste Management Licensing Regulations 1994 (inserted by regulation 3(16) of the Waste Management (Amendment etc) Regulations 1995).

[27] The equivalent provision in Scotland is section 30 of the Civic Government (Scotland) Act 1982 (the 1982 Act).

[28] Section 30(3)(b) of the 1982 Act provides that in Scotland records may alternatively be kept by "the use of a device for storing and processing information."

7.6 Section 2(2) and (3) of the 1964 Act specifies the particulars to be entered in the book as follows:-

In the case of scrap metal received at the place occupied by the dealer

"(a) the description and weight of the scrap metal;

(b) the date and time of the receipt of the scrap metal;

(c) if the scrap metal is received from another person, the full name and address of that person;

(d) the price, if any, payable in respect of the receipt of the scrap metal, if that price has been ascertained at the time when the entry in the book relating to that scrap metal is to be made;

(e) where the last preceding paragraph does not apply, the value of the scrap metal at the time when the entry is to be made as estimated by the dealer;

(f) in the case of scrap metal delivered at the place in question by means of a mechanically propelled vehicle bearing a registration mark (whether the vehicle belongs to the dealer or not), the registration mark borne by the vehicle."

In the case of scrap metal either processed at, or despatched from, the place occupied by the dealer

"(a) the description and weight of the scrap metal;

(h) the date of processing or, as the case may be, despatch of the scrap metal, and, if processed, the process applied;

(I) in the case of scrap metal despatched on sale or exchange, the full name and address of the person to whom the scrap metal is sold or with whom it is exchanged, and the consideration for which it is sold or exchanged;

(j) in the case of scrap metal processed or despatched otherwise than on sale or exchange, the value of the scrap metal immediately before its processing or dispatch as estimated by the dealer."

7.7 As indicated at paragraph 7.4 above, the form of the transfer note required for the purposes of the duty of care is not prescribed in the 1991 Regulations. The purpose of this was to allow each sector of industry to use or adapt its existing systems. This approach enables the scrap metal industry to use the records which dealers are required to keep under the 1964 Act[29] as a basis to fulfil the requirements of the 1991 Regulations.

---

[29] In Scotland the Civic Government(Scotland) Act 1982.

7.8 The only other information required by the 1991 Regulations is:-

    (a) how the material is contained on its transfer (ie whether it is loose or in a container and the kind of any container eg in sacks a skip or drums);

    (b) whether the person transferring the waste is the producer or importer of that material;

    (c) to which category of authorised persons[30] the person transferring the waste and the person receiving the waste belongs; and

    (d) the place of transfer.[31]

7.9 **Annex C** paragraph C.8 provides guidance on the completion of transfer notes.

Keeping a record of transactions

7.10 Section 2(5) of the 1964 Act[32] requires scrap metal dealers to retain the book containing records of dealings for two years beginning from the day on which the last entry was made in the book. This is similar to the requirement under the 1991 Regulations. The 1991 Regulations do not specify which of the two parties must complete the transfer note. It may be completed by either party. However, both parties are required to ensure that it is completed and must sign it and keep a copy.

7.11 Scrap metal dealers may wish, therefore, to assist those who transfer scrap metal to them by providing prepared documentation including the required information which the transferor may then complete and sign. If prepared documentation is provided the transferor should ensure that he is advised by the dealer that the details are correct.

---

[30] If the transfer is to a person for authorised transport purposes, it is necessary to specify which of those purposes.

[31] Regulation 2(2)(a)(iii) of the 1991 Regulations. In the case of scrap metal despatched from the place occupied by the dealer, the transfer note should also record the time of transfer.

[32] In Scotland section 30(4) of the 1982 Act.

Records kept by itinerant traders

7.12 Among other matters, section 3 of the 1964 Act makes provision for dealers who carry on their business as part of the business of an itinerant collector. This sets out the particular record keeping requirements for itinerant collectors, although it remains the case that transaction receipts should be kept for two years. From 1 October 1995 itinerant collectors are subject to the duty of care in the same way as anyone else who deals in scrap metal. Itinerant collectors must therefore ensure that a transfer note is completed and that they retain a copy when they collect or hand on scrap metal. An important exception to this relates to householders' own waste collected from their home (see paragraph 7.15 below).

7.13 In Scotland, the equivalent term for an itinerant collector is an itinerant metal dealer. Provisions covering itinerant metal dealers are contained in sections 32 and 33 of the Civic Government (Scotland) Act 1982 (the 1982 Act). Section 33(3) of the 1982 Act provides that ".. an itinerant metal dealer shall keep a record in respect of each sale to him of metal..." section 33(4) goes on to say "any such records shall be kept by the dealer for a period of 6 months from the date of the sale to which it relates". Therefore, if itinerant metal dealers in Scotland wish to use these record for the purpose of the duty of care provisions they must keep them for two years.

Records kept by motor vehicle dismantlers

7.14 Any motor vehicle dismantler who falls outside the provisions of the 1964 Act because the materials in which he deals are not scrap metal, should follow the guidance on transfer notes provided in paragraphs 1.7-1.9 above and in **Annex C** of this Code of Practice.[33]

Householders

7.15 Householders are **exempt** from the duty of care for their own household waste (**Annex A** paragraph A.9). This means that a transfer note does not need to be completed when a

---

[33] Section 9(1) of the Scrap Metal Dealers Act 1964 defines a scrap metal dealer as follows:- "For the purposes of this Act a person carries on business as a scrap metal dealer if he carries on a business which consists wholly or partly of buying and selling scrap metal, whether the scrap metal sold is in the form in which it was bought or otherwise, other than a business in the course of which scrap metal is not bought except as materials for the manufacture of other articles and is not sold except as a by-product of such manufacture or as surplus materials bought but not required for such manufacture; and `scrap metal dealer' (where that expression is used in this Act otherwise than in a reference to carrying on business as a scrap metal dealer) means a person who (in accordance with the preceding provisions of this subsection) carries on business as a scrap metal dealer."

householder brings his own household waste to a MRS; or when a householder's own scrap metal is collected from his home.

Season tickets

7.16 For regular customers and suppliers there is provision to agree a "season ticket". In other words, one transfer note which will cover multiple transfers over a given period of time. The use of a season ticket is, however, only permissible where the parties involved in the series of transfers do not change and where the description of the waste transferred remains the same (see **Annex C paragraph C.4**).

Transferring and receiving waste

7.17 The guidance provided in **Section 3** (Transfer to the right person) and **Section 4** (Receiving waste) applies to anyone subject to the duty of care, including operators of MRSs. As those **Sections** explain, controlled waste may be transferred only to authorised persons or to persons for authorised transport purposes. A full list of authorised persons is at **Section 3,** paragraph 3.2.

7.18 It is recognised that the scrap metal industry receives its material from diverse sources and that it could take considerable time to verify carriers' or suppliers' credentials at a site which undertakes a great number of transactions each day. The code is not intended to place dealers in the position where they have to seek excessive verification from their suppliers to an extent that in time, discourages them from transferring scrap metal for recovery.

7.19 As **Section 4** explains, checking back when waste is received need not be as thorough as checking forward. There is no explicit requirement for a person receiving waste from a waste carrier to check on whether or not the carrier is registered. In the Departments' view, it is not an offence under section 34(1) of the 1990 Act for a scrap metal dealer[34] to accept waste from an unregistered carrier. However, an offence might have been committed earlier in the chain. For example, a waste producer who transfers waste to someone who is not authorised to accept it. This may be a carrier who was required to be registered but was not. The carrier himself would also be in breach of the carrier registration legislation if he was required to be registered but was not.

---

[34] Or any other person subject to the duty of care.

7.20 However, the first time a carrier delivers waste, the recipient should satisfy himself that he is dealing with someone who is properly registered to transport the waste. In this case it would be reasonable to ask to see either the carrier's registration certificate or official copy certificate, or to request confirmation of the type of exemption under which the carrier is transporting the waste.

Role of the Environment Agency and the Scottish Environment Protection Agency

7.21 **Section 5** (paragraphs 5.10-5.11) explains that, whilst the Agencies do not have a specific duty to enforce the duty of care. they have a major interest in breaches of the duty which might contribute to illegal waste management. The Agencies can, as can anyone else, bring a prosecution if they have evidence that the duty of care has been breached. In their dealings with the scrap metal industry, the Agencies have been asked to have regard to the considerations referred to in paragraph 7.2 above.

SUMMARY CHECKLIST

This section draws together in one place a simple checklist of the main steps that are normally necessary to meet the duty of care. As with the code itself, *this does not mean that completing the steps listed here is all that needs to be done under the duty of care*. The checklist cross-refers to key sections of the code, to the introduction to the code and to the Appendix for fuller guidance. It should be noted that the introduction to the code and the Appendix do not form part of the code of practice itself.

Refer to paragraphs

(a) Is what you have waste? If yes, — Introduction and the Appendix

(b) is it controlled waste? If yes, — 1.2

(c) while you have it, protect and store it properly, — 2.1-2.4

(d) write a proper description of the waste, covering:- — 1.7-1.9

- any problems it poses; — 1.3-1.6 & 1.10

and, *as necessary to others who might deal with it later*, one or more of:-

- the type of premises the waste comes from; — 1.11-1.12

- what the waste is called; — 1.13

- the process that produced the waste; and — 1.14-1.15

- a full analysis; — 1.16-1.17

(e) select someone else to take the waste. They must be an authorised person or authorised for transport purposes. As such they should be one or more of the following and must prove that they are:-

- a registered waste carrier; — 3.6-3.8

- exempt from waste carrier registration; — 3.9-3.12

- a waste manager licensed to accept the waste; — 3.13-3.16

	- exempt from waste management licensing;	3.17-3.18
	- a waste collection authority;	3.5
	- In Scotland, a waste disposal authority operating under the terms of a resolution;	3.20-3.21
	- authorised transport purposes;	3.3
(f)	anyone arranging on behalf of another person for the disposal or recovery of controlled waste must be:-	
	- a registered waste broker or	3.4, 3.19, 4.10
	- an exempt waste broker	3.4
(g)	pack the waste safely when transferring it and keep it in your possession until it is to be transferred;	2.5-2.7
(h)	check the next person's credentials when transferring waste to them;	3.7-3.12, 3.15-3.19
(i)	complete and sign a transfer note;	Annex C, C.3-C.4
(j)	hand over the description and complete a transfer note when transferring the waste;	1.7-1.9 Annex C, C.3-C.4
(k)	keep a copy of the transfer note signed by the person the waste was given to, and a copy of the description, for two years;	Annex C, C.5
(l)	when *receiving* waste, check that the person who hands it over is one of those listed in (e), or the producer of the waste, obtain a description from them, complete a transfer note and keep the documents for two years.	4.2-4.10, 5.4
(m)	whether transferring *or* receiving waste, be alert for any evidence or suspicion that the waste you handle is being dealt with illegally at any stage, in case of doubt question the person involved and if not satisfied, alert the Environment Agency or the Scottish Environment Protection Agency.	5.5-5.11
(n)	supplementary guidance for holders of scrap metal	Section 7

ANNEX A

THE LAW ON THE DUTY OF CARE

What the duty requires

A.1 The duty of care is set out in section 34 of the 1990 Act (a copy is printed at the end of this **Annex**). Those subject to the duty must try to achieve the following four things:-

(a) to prevent any other person committing the offences of depositing, disposing of or recovering controlled waste without a waste management licence[35]; contrary to the conditions of a licence[36]; or in a manner likely to cause environmental pollution or harm to health[37] (It should also be noted that, where a person purports to be carrying on an activity which is exempt from licensing but he fails to comply with the terms and conditions of the exemption, he may be prosecuted under section 33(1) of the 1990 Act for carrying out a licensable activity without a licence.);

(b) to prevent the escape of waste, that is, to contain it;

(c) to ensure that, if the waste is transferred, it goes only to an "authorised person" or to a person for "authorised transport purposes". **A list of authorised persons is provided at paragraph 3.2.** The list of authorised transport purposes is set out in section 34(4) of the 1990 Act which is printed at the end of this **Annex**;

(d) when waste is transferred, to make sure that there is also transferred a written description of the waste, a description good enough to enable each person receiving it to avoid committing any of the offences under (a) above; and to comply with the duty at (b) above to prevent the escape of waste.

A.2 Those subject to the duty must also comply with the 1991 Regulations which require them to keep records and make them available to the Agencies. The 1991 Regulations are additional to the code of practice and are summarised at **Annex C**.

---

[35] Section 33(1)(a) and (b) of the 1990 Act as modified by paragraph 9 of Schedule 4 to the 1994 Regulations.

[36] Section 33(6) of the 1990 Act.

[37] Section 33(1)(c) of the 1990 Act.

A.3 **Failing to observe the duty of care or the 1991 Regulations is a criminal offence.**

Amendments to section 34

A.4 Since section 34 of the 1990 Act was originally brought into force there have been two amendments to it:-

(a) section 33 of the Deregulation and Contracting Out Act 1994 clarifies the transfer note requirements relating to multiple transfers of waste (the "season ticket" arrangements - see also **Annex C** paragraph C.4 and **Section 7** paragraph 7.16). This amendment was made because section 34 of the 1990 Act might have been construed as requiring a separate description to be supplied with each load of waste. Section 33 (4A) of the 1994 Act therefore provides that:-

"(a) a transfer of waste in stages shall be treated as taking place when the first stage of the transfer takes place; and

(b) a series of transfers between the same parties of waste of the same description shall be treated as a single transfer taking place when the first of the transfers in the series takes place".

(b) Paragraph 65 of Schedule 22 of The Environment Act 1995 amends section 34 of the 1990 Act to allow the Secretary of State to make regulations to specify persons as authorised persons under the duty of care as follows:

"(3A) The Secretary of State may by regulations amend subsection (3) above so as to add, whether generally or in such circumstances as may be prescribed in the regulations, any person specified in the regulations, or any description of person so specified, to the persons who are authorised persons for the purposes of subsection (1) (c) above."

Who is under the duty?

A.5 The duty and therefore this code apply to any person who:-

(a) imports, produces or carries controlled waste;

(b) keeps, treats or disposes of controlled waste (a"waste manager"); or

(c) as a broker has control of controlled waste.

A.6 In this code and its **Annexes** all the persons referred to in paragraph A.5 above are referred to as "holders" of waste. Brokers are included in this term for convenience. However, the description in paragraph A.7 below should be borne in mind by users of the code.

A.7 Section 34 of the 1990 Act does not define "broker". However regulation 20 of the 1994 Regulations describes a broker as a person who arranges "for the disposal or recovery of controlled waste on behalf of another person". A broker does not handle waste himself or have it in his own physical possession but he controls what happens to it.

A.8 Employers are responsible for the acts and omissions of their employees. They therefore should provide adequate equipment, training and supervision to ensure that their employees observe the duty of care.

Exemption for householders
A.9 The only exception to the duty is for occupiers of domestic property for "household waste" (see entry for "controlled waste" in the Glossary of terms) from the property. Note that this does *not* exempt:-

(a) a householder disposing of waste that is not from his property (for example, waste from his workplace; or waste from his neighbour's property); or

(b) someone who is not the occupier of the property (for example, a builder carrying out works to a house he does not occupy is subject to the duty for the waste he produces).

Exemption for animal by-products
A.10 Animal waste which is collected and transported in accordance with Schedule 2 to the Animal By-Products Order 1992 is not subject to the duty of care. The 1994 regulations amend the Controlled Waste Regulations 1992 so that this type of waste is not treated as controlled waste for the purposes of the duty of care[38]. Schedule 2 of the Animal By-Products Order

---

[38] Regulation 24(7) of the Waste Management Licensing Regulations 1994 amends regulation 7 of the Controlled Waste Regulations 1992 (S.I. 1992 No.588).

contains a system of control on the transfer of waste which achieves broadly the same objective as the duty of care.

What each waste holder has to do

A.11 A waste holder is not responsible for ensuring that all the aims of the duty of care (listed in paragraph A.1) are fulfilled. A holder is only expected to take measures that are:-

    (a) reasonable in the circumstances, *and*

    (b) applicable to him in his capacity.

A.12 The *circumstances* that affect what is reasonable include:-

    (a) what the waste is;

    (b) the dangers it presents in handling and treatment;

    (c) how it is dealt with; and

    (d) what the holder might reasonably be expected to know or foresee.

A.13 The *capacity* of the holder is who he is, how much control he has over what happens to the waste and in particular what his connection with the waste is. Different measures will be reasonable depending on whether his connection with the waste is as an importer, producer, carrier, keeper, treater, disposer, dealer or broker.

A.14 Waste holders can be responsible only for waste which is at some time under their control. It is not a breach of the duty to fail to take steps to prevent someone else from mishandling any other waste. However, a holder's responsibility for waste which he at any stage controls extends to what happens to it at other times, insofar as he knows or might reasonably foresee. Further guidance on the reasonable limits of responsibilities is in **Annex B**.

Offences and penalties

A.15 Breach of the duty of care is a criminal offence. It is an offence irrespective of whether or not there has been any other breach of the law or any consequent environmental pollution or harm to human health. The offence is punishable by a fine of up to £5,000 on summary conviction or an unlimited fine on conviction on indictment.

The Code of Practice

A.16 This code of practice has statutory standing. The Secretary of State is obliged by section 34 of the 1990 Act to issue practical guidance on how to discharge the duty. This code constitutes that guidance. The code is admissible in evidence in court. Section 34 (10) of the 1990 Act provides that if a provision of the code appears to the court to be relevant to any question arising in the proceedings, it must be taken into account in determining that question. The code may therefore be used by a Court in deciding whether or not an accused has complied with his duty of care.

Duty of care etc. as respects waste

Duty of care etc. as respects waste.

34.-(1) Subject to subsection (2) below, it shall be the duty of any person who imports, produces, carries, keeps treats or disposes of controlled waste or, as a broker, has control of such waste, to take all such measures applicable to him in that capacity as are reasonable in the circumstances -

    (a) to prevent any contravention by any other person of section 33 above;

    (b) to prevent the escape of waste from his control or that of any other person; and

    (c) on the transfer of the waste to secure -

        (I) that the transfer is only to an authorised person or to a person for authorised transport purposes; and

        (ii) that there is transferred such a written description of the waste as will enable other persons to avoid a contravention of that section and to comply with the duty under this subsection as respects the escape of waste.

(2) The duty imposed by subsection (1) above does not apply to an occupier of domestic property as respects the household waste produced on the property.

(3) The following are authorised persons for the purpose of subsection (1)(c) above -

    (a) any authority which is a waste collection authority for the purposes of this Part;

1974 c.40

    (b) any person who is the holder of a waste management licence under section 35 below or of a disposal licence under section 5 of the Control of Pollution Act 1974;

1989 c14

    (c) any person to whom section 33(1) above does not apply by virtue of regulations under subsection (3) of that section;

    (d) any person registered as a carrier of controlled waste under section 2 of the Control of Pollution (Amendment) Act 1989;

    (e) any person who is not required to be so registered by virtue of regulations under section 1(3) of that Act; and

    (f) a waste disposal authority in Scotland.

(3A) The Secretary of State may by regulations amend subsection (3) above so as to add, whether generally or in such circumstances as may be prescribed in the regulations, any person specified in the regulations, or any description of person so specified, to the persons who are authorised persons for the purposes of subsection (1) (c) above.

(4) The following are authorised transport purposes for the purposes of subsection (1)(c) above -

(a) the transport of controlled waste within the same premises between different places in those premises;

(b) the transport to a place in Great Britain of controlled waste which has been brought from a country or territory outside Great Britain not having been landed in Great Britain until it arrives at that place; and

(c) the transport by air or sea of controlled waste from a place in Great Britain to a place outside Great Britain;

and "transport" has the same meaning in this subsection as in the Control of Pollution (Amendment) Act 1989.

(4A) For the purposes of subsection (1)(c)(ii) above -

(a) a transfer of waste in stages shall be treated as taking place when the first stage of the transfer takes place, and

(b) a series of transfers between the same parties of waste of the same description shall be treated as a single transfer taking place when the first of the transfers in the series takes place.

(5) The Secretary of State may, by regulations, make provision imposing requirements on any person who is subject to the duty imposed by subsection (1) above as respects the making and retention of documents and the furnishing of documents or copies of documents.

(6) Any person who fails to comply with the duty imposed by subsection (1) above or with any requirement imposed under subsection (5) above shall be liable -

(a) on summary conviction, to a fine not exceeding the statutory maximum; and

(b) on conviction on indictment, to a fine.

(7) The Secretary of State shall, after consultation with such persons or bodies as appear to him representative of the interests concerned, prepare and issue a code of practice for the purpose of providing to persons practical guidance on how to discharge the duty imposed on them by subsection (1) above.

(8) The Secretary of State may from time to time revise a code of practice issued under subsection (7) above by revoking, amending or adding to the provisions of the code.

(9) The code of practice prepared in pursuance of subsection (7) above shall be laid before both Houses of Parliament.

(10) A code of practice issued under subsection (7) above shall be admissible in evidence and if any provision of such a code appears to the court to be relevant to any question arising in the proceedings it shall be taken into account in determining that question.

(11) Different codes of practice may be prepared and issued under subsection (7) above for different areas.

ANNEX B

RESPONSIBILITIES UNDER THE DUTY OF CARE

B.1 The main **Sections** of the code address all waste holders who are subject to the duty of care. In law all the responsibilities for waste under the duty are spread among all those who hold that waste at any stage. But responsibilities are not spread evenly. Some holders will have greater or less responsibility for some aspects of the duty, according to their connection with the waste (see **Annex A** paragraphs A.11-A.14). This **Annex** offers guidance for each category of waste holder on their particular responsibilities. This **Annex** is neither comprehensive nor self-contained guidance for particular categories of waste holder. It only draws out questions of the allocation of responsibility. All waste holders should follow the guidance in the main sections of the code of practice.

Waste producers

B.2 The duty of care applies to a person who "produces" waste. Section 34 of the 1990 Act does not define this term. However, in relation to the definition of waste, regulation 1(3) of the 1994 Regulations provides that "'producer' means anyone whose activities produce [Directive] waste or who carries out preprocessing, mixing or other operations resulting in a change in its nature or composition". This definition is also reflected in paragraph 88 of Schedule 22 of the Environment Act 1995.

B.3 Waste producers are solely responsible for the care of their waste while they hold it. Waste producers are normally best placed to know what their waste is and to choose the disposal, treatment or recovery method, if necessary with expert help and advice. They bear the main responsibility for ensuring that the description of the waste which leaves them is accurate and contains all the information necessary for safe handling, disposal, treatment or recovery. If they also select a final disposal, treatment or recovery destination then they share with the waste manager of that destination responsibility for ensuring that the waste falls within the terms of any licence or exemption relevant to that final destination.

B.4 Producers bear the main responsibility for packing waste to prevent its escape in transit. Waste leaving producers should be packed in a way that subsequent holders can rely on.

B.5 Using a registered or exempt carrier does not necessarily let a producer out of all responsibility for checking the later stages of the disposal of his waste. The producer and the disposer may sometimes make all the arrangements for the disposal or recovery of waste, and then contract with a carrier simply to convey the waste from one to the other. Such a case is very little different in practice from that where there is no intermediate carrier involved. If a producer arranges disposal or recovery then he should exercise the same care in selecting the disposer or recoverer as if he were delivering the waste himself.

B.6 It is not possible to draw a line at the gate of producers' premises and say that their responsibility for waste ends there. A producer is responsible according to what he knows or should have foreseen. So if he hands waste to a carrier not only should it be properly packed when transferred, but the producer should take account of anything he sees or learns about the way in which the carrier is subsequently handling it. The producer would not be expected to follow the carrier but he should be able to see whether the waste is loaded securely for transport when it leaves, and he may come to learn or suspect that it is not ending up at a legitimate destination. A producer may notice a carrier's lorries returning empty for further loads in a shorter time than they could possibly have taken to reach and return from the nearest lawful disposal site; or a producer may notice his carrier apparently engaged in the unlawful dumping of someone else's waste. These would be grounds for suspecting illegal disposal of his own waste. The same reasoning applies when a producer makes arrangements with a waste manager for the disposal, treatment or recovery of waste. The producer shares the blame for illegal treatment of his waste if he ignores evidence of mistreatment. A producer should act on knowledge to stop the illegal handling of waste (see paragraphs 5.5-5.12).

Waste importers
B.7 Revised legislative requirements mean that waste importers and exporters face stringent controls and in some cases, prohibitions on transfrontier shipments of waste. Consent to movements of waste must be obtained from the competent authorities involved and full details about each shipment must be given on a consignment note (**Annex D** vi. D.21 -D.22).

Waste carriers
B.8 A waste carrier is responsible for the adequacy of packaging while waste is under his control. He should not rely totally on how it is packed or handed over by the previous holder. He should at least look at how it is contained to ensure that it is not obviously about to escape. His own handling in transferring and transporting waste should take account of how it is packed, or if necessary he may need to repack it or have it repacked to withstand handling.

B.9 A waste carrier would not normally be expected to take particular measures to provide a new description of the waste he carried unless he altered it in some way. The description would normally be provided by the waste producer, unless the producer is a householder not subject to the duty, in which case the person first taking waste from the householder would have to ensure that a description was furnished to the next holder. When accepting any waste a carrier should make at least a quick visual inspection to see that it appears to match the description, but he need not analyse the waste unless there is reason to suspect an anomaly or the waste is to be treated in some way not foreseen by the producer.

B.10 If a carrier or any other intermediate holder does alter waste in any way, by mixing, treating or repacking it, he will need to consider whether a new description is necessary. The description received may continue to serve for waste that is merely compacted or mixed with similar waste, but if it has deteriorated or decomposed or if it has been altered in any way that matters for handling and disposal then a new description should be written.

B.11 Where there is a contract between the producer and the waste manager and a carrier is contracted solely to provide transport, he may rely on the producer checking the scope of the licence or exemption of the waste manager to whom the waste is delivered. In other cases the carrier should check this himself.

Waste brokers

B.12 A waste broker arranging the transfer of waste between a producer and a waste manager, to such an extent that he controls what happens to the waste, is taking responsibility for the legality of the arrangement. He should ensure that he is as well informed about the nature of the waste as if he were discharging the responsibilities of both producer and waste manager. He is as responsible as either for ensuring that a correct and adequate description is transferred, that the waste is within the scope of any waste licence or exemption, that it is carried only by a registered or exempt carrier and that documentation is properly completed. As he does not directly handle the waste, he cannot be held responsible for its packaging. However, he may be in a position to advise producers on appropriate methods of packaging or containment if he has greater knowledge of how the waste will be handled once it is removed from the producers premises. He should also undertake the same level of checks after transfer, and the same action on any cause for suspicion, as a waste holder.

Waste managers

B.13 Waste managers, like waste carriers, should normally be able to rely on the description of waste supplied to them. However in disposing, treating or recovering waste they are in a stronger position to notice discrepancies between the description and the waste and therefore bear a greater responsibility for checking descriptions of waste they receive. Sample checks on the composition of waste received would be good practice.

B.14 Waste managers are also in a good position to notice that waste has not been properly dealt with or falsely documented before it reaches them . For example, a waste manager may receive waste which is documented as coming directly from one producer but which shows signs of having been mixed or treated at some intermediate stage. A manager is to some extent responsible for following up evidence of previous misconduct just as he is for subsequent mismanagement of waste, that is to the extent that he knew or should have foreseen it and to the extent that he can control what happens.

ANNEX C

REGULATIONS ON KEEPING RECORDS

C.1 The 1991 Regulations[39] are made under section 34(5) of the 1990 Act and require all those subject to the duty to make records of waste they receive and consign, to keep the records and to make them available to the Environment Agency or to the Scottish Environment Protection Agency.

C.2 The 1991 Regulations require each party to any transfer to keep a copy of the description which is transferred. An individual holder might transfer onward the description of the waste that he received unchanged in which case it would be advisable for the sake of clarity to endorse the description for onward transfer to the effect that the waste was sent onwards as received. If a different description of waste is transferred onwards, whether or not this reflects any change in the nature or composition of the waste, then copies of both descriptions must be made. The holder making the copy need not be the author of the description, which will often be written only by the producer or broker and reused unchanged by each subsequent holder.

C.3 The Regulations also require the parties to complete, sign and keep a transfer note. The transfer note contains information about the waste and about the parties to the transfer.

Season ticket provisions

C.4 While all transfers of waste must be documented, the 1991 Regulations do not require *each* individual transfer to be separately documented. Where a series of transfers of waste of the same description is being made between the same parties, provision is made for the parties to agree a "season ticket" - ie one transfer note covering a series of transfers (see paragraph A.4 above). A season ticket might be used, for example, for the weekly or daily collections of waste from shops or commercial premises, or the removal of a large heap of waste by multiple lorry trips. In the Departments' view, however, a season ticket should not extend for a period of more than 12 months from the date on which the first of the transfers subject to the arrangements takes place.

---

[39] The Environmental Protection (Duty of Care) Regulations 1991 (S.I. 1991 No.2839).

Duty of care: Controlled waste transfer note

C.5 The 1991 Regulations require these records (both the descriptions and the transfer notes) to be kept for at least two years. Holders (which includes where relevant, brokers) must provide copies of these records if requested by the Agencies.

C.6 One purpose of documentation is to create an information source of use to other holders. It is open to holders (including where relevant, brokers) to ask each other for details from records, especially to check what happened to waste after it was consigned. A holder or broker might draw conclusions and alert the Agencies to any suspected breach of the duty if such a request were refused.

Format of the record

C.7 There is no compulsory form for keeping these records. It is recognised that a number of holders already keep records of waste in a manner that meets the requirements of the 1991 Regulations with little or no further adaptation. For example, scrap metal dealers may adapt the system of books they are obliged to keep under the Scrap Metal Dealers Act 1964 or in Scotland the Civic Government (Scotland) Act 1982 (see **Section 7** paragraphs 7.5-7.13). The consignment note for special waste can be properly completed so as to fulfil the duty of care requirements (**Annex D** paragraphs D.16- D.18). Similarly, the consignment note system for transfrontier shipments of waste which involves the use of movement and tracking forms should, if properly completed satisfy the duty of care arrangements (**Annex D** paragraphs D.21-D.22).

C.8 A suggested standard form for voluntary use is included in this **Annex**. It should be noted that the boxes on the model form are consistent with the requirements of section 34(1) of the 1990 Act and with the requirements of the 1991 Regulations . This means that where existing documentation systems are adapted to meet the transfer note requirements of the duty of care, all of this information should be included.

C.9 **Breach of any provision of the 1991 Regulations is an offence.**

Duty of Care: Controlled Waste Transfer Note

Section A - Description of Waste

1. Please describe the waste being transferred:

2. How is the waste contained?

   Loose ☐   Sacks ☐   Skip ☐   Drum ☐   Other ☐ → please describe:

3. What is the quantity of waste (number of sacks, weight etc):

Section B - Current holder of the waste (Transferor)

1. Full Name (BLOCK CAPITALS):

2. Name and address of Company:

3. Which of the following are you? (Please ✓ one or more boxes)

   producer of the waste ☐   holder of waste disposal or waste management licence ☐ → Licence number: / Issued by:

   importer of the waste ☐   exempt from requirement to have a waste disposal or waste management licence ☐ → Give reason:

   waste collection authority ☐   registered waste carrier ☐ → Registration number: / Issued by:

   waste disposal authority (Scotland only) ☐   exempt from requirement to register ☐ → Give reason:

Section C - Person collecting the waste (Transferee)

1. Full Name (BLOCK CAPITALS):

2. Name and address of Company:

3. Which of the following are you? (Please ✓ one or more boxes)

   authorised for transport purposes ☐ → Specify which of those purposes:

   waste collection authority ☐   holder of waste disposal or waste management licence ☐ → Licence number: / Issued by:

   waste disposal authority (Scotland only) ☐   exempt from requirement to have a waste management licence ☐ → Give reason:

   registered waste carrier ☐ → Registration number: / Issued by:

   exempt from requirement to register ☐ → Give reason:

Section D

1. Address of place of transfer/collection point:

2. Date of transfer: 3. Time(s) of transfer (for multiple consignments, give 'between' dates):

4. Name and address of broker who arranged this waste transfer (if applicable):

Transferor	Transferee
5. Signed:	Signed:
Full name: (BLOCK CAPITALS)	Full Name: (BLOCK CAPITALS)
Representing:	Representing:

FED 0443 (02/96 DDP)

ANNEX D

OTHER LEGAL CONTROLS

D.1 This code offers guidance on the discharge of a waste holder's duty of care under section 34 of the 1990 Act. Holders are also subject to other statutory requirements, some of the most important of which are set out here.

i. Waste management licensing

D.2 In *England, Wales* and *Scotland* a new system of waste management licensing under Part II of the 1990 Act came into force on 1 May 1994 and replaced the system of licensing set up under the Control of Pollution Act 1974. (The new system applied to the recovery of scrap metal and waste motor vehicle dismantling from 1 April 1995). The new system is the main means by which the Government's obligations under the amended EC Framework Directive on waste are fulfilled.

D.3 Under the 1990 Act *waste management licences* are required to authorise the:-

(a) deposit of controlled waste in or on land;

(b) the disposal or recovery of controlled waste (ie a disposal operation or a recovery operation);

(c) the use of certain mobile plant to dispose of or recover controlled waste.

D.4 Exemptions from the requirement to have a waste management licence for any of the activities listed above are to be found in regulations 16 and 17 of and Schedule 3 to the 1994 Regulations[40].

---

[40] As amended by The Waste Management Licensing (Amendment Etc.) Regulations 1995 (S.I. 1995 No.288) and by The Waste Management Licensing (Amendment No 2) Regulations 1995 (S.I. 1995 No.1950).

New controls

D.5 The Environment Act 1995 transfers waste regulation functions in *England and Wales* from waste regulation authorities to the Environment Agency, and in *Scotland* to the Scottish Environment Protection Agency on 1 April 1996.

D.6 Paragraph 65 of Schedule 22 of the Environment Act 1995 also provides that additions may be made to the list of "authorised persons" (listed in **Section 3** paragraph 3.2), for the purpose of the duty of care.

ii. The registration of waste carriers

D.7 Subject to certain provisions, section 1(1) of the Control of Pollution (Amendment) Act 1989 ("the 1989 Act") makes it an offence to transport controlled waste without being registered as a waste carrier. The requirement to register applies to any person who transports controlled waste which that person has not produced themselves, to or from any place in Great Britain in the course of any business of his or otherwise with a view to profit. For this purpose, "transport" includes the transport of waste by road, rail, air, sea or inland waterway. It should be noted that construction (which includes improvement, repair or alteration) and demolition contractors would have to be registered as carriers if they wished to transport such waste, even if they had produced it themselves, and the transport of such waste when not registered would constitute a breach of the duty of care. The Controlled Waste (Registration of Carriers and Seizure of Vehicles) Regulations 1991[41] require the Agencies to establish and maintain a register of waste carriers; and set out the basis on which the registration system operates. Guidance on registration is provided in DOE Circular 11/91[42].

D.8 Paragraph 12 of Schedule 4 to the 1994 Regulations requires any waste carrier exempted from the waste carrier registration requirements (ie charities, voluntary organisations, waste collection/ British Rail subsidiaries and the Agencies) to register as an exempt carrier with the Environment Agency or with the Scottish Environment Protection Agency.

---

[41] S.I. 1991 No.1624 as amended by S.I. 1992 No.588 and S.I. 1994 No.1056.

[42] Welsh Office Circular 34/91 and The Scottish Office Environment Department Circular 18/91.

D.9 Anyone subject to the duty of care must ensure that, if waste is transferred, it is transferred only to an authorised person or to a person for authorised transport purposes. Among those who are "authorised persons" are:-

(a) any person registered with the Agencies as a carrier of controlled waste; and

(b) any person who is exempt from registration by virtue of regulations made under section 1(3) of the 1989 Act. The exemptions at present in force are set out in regulation 2 of the Controlled Waste (Registration of Carriers and Seizure of Vehicles) Regulations 1991.

D.10 It is not an offence under section 1(1) the 1989 Act to transport controlled waste without being registered in the circumstances set out in (a)-(c) below. The following are also "authorised transport purposes" for the duty of care:

(a) the transport of controlled waste between different places within the same premises;

(b) the transport to a place in Great Britain of controlled waste which has been brought from a country or territory outside Great Britain and is not landed in Great Britain until it arrives at that place. This means that the requirement to register as a carrier applies only from the point at which imported waste is landed in Great Britain; and

(c) the transport by air or sea of controlled waste being exported from Great Britain.

iii. The registration of waste brokers

D.11 Subject to various provisions, regulation 20 of the 1994 Regulations makes it an offence for anyone to arrange on behalf of another person for the disposal or recovery of controlled waste if he is not registered as a broker. Anyone subject to the duty of care must ensure that, insofar as they use a broker when transferring waste, they use a registered broker or one exempt from the registration requirements.

D.12 The Transfrontier Shipment of Waste Regulations 1994 (S.I. 1994 No.1137) require waste brokers making international arrangements for the disposal or recovery of controlled waste within Great Britain or abroad to be registered.

D.13 Registered waste carriers transporting waste to or from any place in Great Britain as part of the arrangement do not need to be registered as a broker.

D.14 Paragraph 12 of Schedule 4 to the 1994 Regulations requires any undertaking which acts as a waste broker or dealer, is exempted under regulation 20 (ie a charity, voluntary organisation, waste collection/disposal/regulation authority) and possesses no other relevant permit for carrying on this activity, to register with the Agencies.

D.15 Anyone subject to the duty of care must ensure that, if waste is transferred, it is transferred only to an authorised person or to a person for authorised transport purposes. Authorised persons are discussed and listed in **Section 3** paragraph 3.2 and at **Annex A**.

iv. Special waste

D.16 Certain particularly difficult or dangerous wastes ("special wastes") are subject to additional requirements in The Control of Pollution (Special Waste) Regulations 1980 (S.I. 1980 No.1709) made under the Control of Pollution Act 1974. The 1980 Regulations will be replaced by new Special Waste Regulations made under section 62 of the 1990 Act. The new Regulations are planned to come into force in 1996.

D.17 Special waste is subject to the duty of care, including the guidance in the code of practice and the requirements of the 1991 Regulations, in the same way as other controlled waste. Compliance with the duty of care does not in any way discharge the need also to comply with the Special Waste Regulations.

D.18 The consignment note for special waste can be properly completed so as to fulfil the duty of care requirements; a separate transfer note is not then required.

v. Road transport of dangerous substances

D.19 Waste holders have obligations in respect of the regulations and associated codes of practice concerned with the transport of dangerous substances. For national transport the primary regulations are:-

> (a) The Road Traffic (Carriage of Dangerous Substances in Road Tankers and Tank Containers) Regulations 1992 (S.I. 1992 No.743 as amended by S.I. 1992 No.1213, S.I. 1993 No.1746 and S.I. 1994 No.669;

> (b) The Carriage of Dangerous Goods by Road and Rail (Classification, Packaging and Labelling) Regulations 1994 (S.I. 1994 No.669);

(c) The Road Traffic (Carriage of Dangerous Substances in Packages etc.) Regulations 1992 (S.I. 1992 No.742 as amended by S.I. 1992 No.1213, S.I. 1993 No.1746 and S.I. 1994 No.669);

(d) The Road Traffic (Training of Drivers of Vehicles Carrying Dangerous Goods) Regulations 1992 (S.I. 1992 No.744 as amended by S.I. 1992 No.1213, S.I. 1993 No.1122, S.I. 1993 No.1746, S.I. 1994 No.397 and S.I. 1994 No.669);

(e) The Dangerous Substances in Harbour Areas Regulations 1987 (S.I. 1987 No.37).

D.20 The regulations listed above are supplemented by the Approved Carriage List - Information approved for the classification, packaging and labelling of dangerous goods for carriage by road and rail (ISBN 0 7176 0745 3, available from Health and Safety Executive (HSE) Books.) The regulations are also supplemented by the Approved Methods for the Classification and Packaging of Dangerous Goods for Carriage by Road and Rail (ISBN 0 7176 0744 5, available from HSE Books).

vi International waste transfers

D.21 Waste importers or exporters must act according to the requirements of the EC Waste Shipments Regulation (259/93/EEC) and the associated Transfrontier Shipment of Waste Regulations 1994 (S.I. 1994 No 1137).

D.22 The EC Regulation requires pre-notification of and consent to movements of waste, with the details of each shipment set out on a consignment note. (Non-hazardous waste moving for recovery does not need to be pre-notified). The consignment note adopted by the EC is the OECD notification and movement/tracking form. It is generally agreed in the UK that copies of both documents should accompany each shipment. These two documents together should satisfy the duty of care requirements. Forms are issued by UK/EC competent authorities of dispatch or destination, except for imports of waste for recovery into the EC from an OECD country where they are issued by the competent authority of dispatch in that country.

vii Health and safety

D.23 Waste holders have a duty to ensure, so far as is reasonably practicable, the health and safety of their employees and other persons who may be affected by their actions in connection with the use, handling, storage or transport of waste, by virtue of sections 2 and 3 of the Health and Safety at Work etc Act 1974.

D.24 Holders also have specific duties under health and safety regulations, including the Control of Substances Hazardous to Health Regulations 1994 (S.I. 1994 No.3246), and in particular the requirement to carry out an assessment of the risks of their activities as required by the Management of Health and Safety at Work Regulations 1992 (S.I. 1992 No 2051).

ANNEX E
GLOSSARY OF TERMS USED IN THIS CODE OF PRACTICE

The 1995 Act: The Environment Act 1995.

The 1990 Act: the Environmental Protection Act 1990.

The 1989 Act: the Control of Pollution (Amendment) Act 1989.

The 1982 Act: The Civic Government (Scotland) Act 1982.

The 1964 Act: The Scrap Metal Dealers Act 1964.

The Agencies: The Environment Agency and the Scottish Environment Protection Agency. These are the waste regulation authorities in Great Britain from 1 April 1996.

Broker: a person who arranges for the disposal or recovery of controlled waste on behalf of another. Such arrangements will include those for the transfer of waste. He does not handle waste himself or have it in his own physical possession, but he controls what happens to it. However, for convenience, brokers are included within the term "waste holder" as it is used in the code of practice.

A registered broker is registered with a waste regulation authority in accordance with regulation 20 of the Waste Management Licensing Regulations 1994.

An exempt broker is a waste broker who by virtue of regulation 20(2), (3) or (4) of the Waste Management Licensing Regulations 1994 is not required to be registered under regulation 20.

Carrier: a person who transports controlled waste, within Great Britain, including journeys into and out of Great Britain.

A registered carrier is registered with a waste regulation authority under the 1989 Act.

An exempt carrier is a waste carrier who is not required to register under the 1989 Act because:-

(a) he is exempt from registration by virtue of regulation 2 of the Controlled Waste (Registration of Carriers and Seizure of Vehicles) Regulations 1991; or

(b) he is transporting waste in the circumstances set out in section 1(2) of the 1989 Act. These circumstances are also "authorised transport purposes" as defined in section 34(4) of the 1990 Act.

Controlled waste: as defined in section 30 of the Control of Pollution Act 1974, section 75 of the 1990 Act and the Controlled Waste Regulations 1992 (as amended). That is, household, commercial and industrial waste. Paragraphs 9(2) and 10(3) to Schedule 4 of the Waste Management Licensing Regulations 1994 provides that any reference to "waste" in Part I of the 1974 Act or Part II of the 1990 Act includes a reference to Directive Waste.

Dealer: a dealer in controlled waste acquires waste and sells it on. He may be a holder of the waste, or he may (as a broker does) make arrangements for its transfer without holding it.

A registered dealer is registered with a waste regulation authority in accordance with regulation 20 of the Waste Management Licensing Regulations 1994.

A registered dealer in scrap metal: A scrap metal dealer registered with the local authority in accordance with the Scrap Metal Dealers Act.1964. For the purposes of that Act a scrap metal dealer is defined as a person who "carries on a business which consists wholly or partly of buying and selling scrap metal ...."

Directive waste: as defined in regulation 1(3) of the Waste Management Licensing Regulations 1994. Directive waste is subject to control as household, industrial or commercial waste (ie as "controlled waste").

Disposal operation: "disposal" as defined in Regulation 1(3) of the Waste Management Licensing Regulations 1994. A list of disposal operations is set out at Part III of Schedule 4 to those Regulations.

The Framework Directive on waste: Council Directive 75/442/EEC as amended by Directives 91/156/EEC and 91/692/EEC. The Framework Directive provides a common European definition of waste; it requires the control through licensing of waste recovery and disposal operations; and defines the circumstances under which exemptions from licensing may be made.

Holder: means a person who imports, produces, carries, keeps, treats, or disposes of controlled waste or, as a broker, has control of it.

Licensed waste manager is one in possession of a licence under section 35 of the 1990 Act or, for a transitional period, under section 5 of the Control of Pollution Act 1974.

An exempt waste manager is one exempted from licensing by regulations under section 33(3) of the 1990 Act.

Metal recycling site (MRS): a site for the recovery scrap metal or waste motor vehicle dismantling. The site may either be subject to waste management licensing or operated under the terms and conditions of an exemption from licensing.

Producer: means anyone whose activities produce [Directive] waste or who carries out preprocessing, mixing or other operations resulting in a change in its nature or composition (regulation 1(3) of the Waste Management Licensing Regulations 1994 (S.I. 1994 No.1056)).

Recovery operation: "recovery" as defined in regulation 1(3) of the Waste Management Licensing Regulations 1994. A list of recovery operations is set out in Part 1V of Schedule 4 to the Waste Management Licensing Regulations 1994.

The 1991 Regulations: The Environmental Protection (Duty of Care) Regulations 1991 (S.I. 1991 No 2839).

The 1994 Regulations: The Waste Management Licensing Regulations 1994 (S.I. 1994 No 1056).

The 1995 Regulations: The Waste Management Licensing (Amendment etc) Regulations 1995 (S.I. 1995 No 288).

Transferee: The person to whom waste is transferred.

Transferor: The person transferring the waste to another person.

Waste collection authority: local authority responsible for collecting waste; as defined in section 30(3) of the 1990 Act.

Waste disposal authority: as defined in section 30(2) of the 1990 Act. That is a local authority responsible for arranging the disposal of publicly collected waste. **In Scotland,** a waste disposal authority may itself act as a waste manager.

Waste regulation authority: as defined in section 30(1) of the 1990 Act. The authority formerly charged with the issue of waste management licences under section 35 of the 1990 Act. Section 30(1) of the 1990 Act originally designated waste regulation as a local authority function. The 1995 Act transfers on 1 April 1996, responsibility for waste regulation to two new Agencies - the Environment Agency in England and Wales and the Scottish Environment Protection Agency in Scotland. Section 30(1) of the 1990 Act is amended accordingly.

APPENDIX

THE DEFINITION OF WASTE

The following is re-produced from DOE Circular 11/94[43]. The highlighted paragraph references are references to paragraphs of DOE Circular 11/94. This Appendix is not part of the code of practice.

"**SUMMARY**

What Is Waste?
2.53 The following paragraphs are intended to provide a helpful summary of the Departments' guidance on the definition of waste; and to draw attention to some of the main questions which should be addressed in reaching a view on whether a particular substance or object is waste. However, the Departments caution against reaching a view on whether any particular substance or object is waste until all of the relevant issues have been considered. The main questions are:-

(a) does the substance or object fall into one of the categories set out in Part II of Schedule 4 to the Regulations; **and**

(b) if so, has it been discarded by its holder, does he have any intention of discarding it or is he required to discard it (**paragraphs 2.18 and 2.19**)?

When Is A Substance Or Object Discarded?
2.54 Waste appears to be perceived in the Directive as posing a threat to human health or the environment which is different from the threat posed by substances or objects which are not waste. This threat arises from the particular propensity of waste to be disposed of or recovered in ways which are potentially harmful to human health or the environment and from the fact that the producers of the substances or objects concerned may no longer have the self interest necessary to ensure the provision of appropriate safeguards. The purpose of the Directive, therefore, is to treat as waste those substances or objects which fall out of the normal commercial cycle or out of the chain of utility (**paragraph 2.14**). To determine whether a substance or object has been discarded the following question should be asked:-

- has the substance or object been discarded so that it is no longer part of the normal commercial cycle or chain of utility?

2.55 An answer of "no" to this question should provide a reasonable indication that the substance or object concerned is not waste (**paragraphs 2.20 and 2.21**). A substance or object should **not** be regarded as waste:-

(a) solely on the grounds that it falls into one of the categories listed in Part II of Schedule 4 to the Regulations (**paragraph 2.19**);

(b) solely on the grounds that it has been consigned to a recovery operation listed in Part IV of Schedule 4 to the Regulations (**paragraphs 2.25 and 2.26**);

---

[43] Welsh Office Circular 26/94 and The Scottish Office Environment Department Circular 10/94.

(c) if it is sold or given away and can be used in its present form (albeit after repair) or in the same way as any other raw material without being subjected to *a specialised recovery operation* (**paragraphs 2.28(a), 2.30-2.31, 2.32, 2.33(a), 2.33(b)(I), 2.35 and 2.36**);

(d) if its producer puts it to beneficial use (**paragraph 2.40**); or

(e) solely on the grounds that its producer would be unlikely to seek a substitute for it if it ceased to become available to him as, say, a by-product (**paragraph 2.42**).

2.56 A substance or object **should be** regarded as waste if it falls into one of the categories listed in Part II of Schedule 4 to the Regulations and:-

(a) it is consigned to a disposal operation listed in Part III of Schedule 4 to the Regulations (**paragraph 2.33**);

(b) it can be used only after it has been consigned to *a specialised recovery operation* (**paragraphs 2.28(b), 2.30-2.31 and 2.33(c)**);

(c) the holder pays someone to provide him with a service, and that service is the collection [and taking away] of a substance or object which the holder does not want and wishes to get rid of (**paragraph 2.33(d)**);

(d) the purpose of any [beneficial] use is wholly or mainly to relieve the holder of the burden of disposing of it and the user would be unlikely to seek a substitute for it if it ceased to become available to him as, say, a by-product (**paragraphs 2.37, 2.38, 2.41 and 2.42**);

(e) it is discarded or otherwise dealt with as if it were waste (**paragraphs 2.33(b)(ii) and 2.52**); or

(f) it is abandoned or dumped (**paragraph 2.24**).

Can Waste Cease To Be Waste?
2.57 A substance or object which is waste **does not** cease to be waste as soon as it is transferred for collection, transport, storage, *specialised recovery* or disposal; or as soon as it reaches a *specialised recovery establishment or undertaking*. A substance or object which is waste, and is not fit for use in its present form or in the same way as any other raw material, may cease to be waste when it has been recovered within the meaning of the Directive (**paragraphs 2.46-2.49**).

2.58 The recovery of waste occurs when its processing produces a material of sufficient beneficial use to eliminate or sufficiently diminish the threat posed by the original production of the waste. This will generally take place when the recovered material can be used as a raw material in the same way as raw materials of non-waste origin by a natural or legal person other than *a specialised recovery establishment or undertaking* (**paragraphs 2.47-2.49**).

2.59 In a few cases, a change of intention by the person to whom waste has been transferred may be sufficient to result in the substance or object concerned ceasing to be waste. This is likely to occur only where it transpires that the substance or object which has been transferred as waste is in fact fit for use in its present form (albeit after repair) or in the same way as any other raw material without being subjected to *a specialised recovery operation* (**paragraph 2.50**)."